Earth

by Steven L. Kipp

Content Consultants:
Rod Nerdahl
Program Director, Minneapolis Planetarium

Diane Kane
Space Center Houston

Bridgestone Books

an imprint of Capstone Press

Bridgestone Books are published by Capstone Press
818 North Willow Street, Mankato, Minnesota 56001
http://www.capstone-press.com

Library of Congress Cataloging-in-Publication Data
Kipp, Steven L.
    Earth/by Steven L. Kipp.
    p. cm.--(The galaxy)
    Summary: Describes the surface features, interior, atmosphere,
magnetic field, and single satellite of the earth.
    ISBN 1-56065-606-9
    1. Earth--Juvenile literature. [1. Earth.] I. Title.
II. Series: Kipp, Steven L.   Galaxy.

QB631.4.K57   1998
550--dc21

                                                97-6917
                                                  CIP
                                                   AC

Photo credits
Capstone Press, 13
International Stock/Chad Ehlers, 10; Warren Faidley, 12; Michael Ventura, 16
NASA, cover, 6, 8, 18, 20
Unicorn/Dinodia, 14; Arthur Gurmankin, 19

Table of Contents

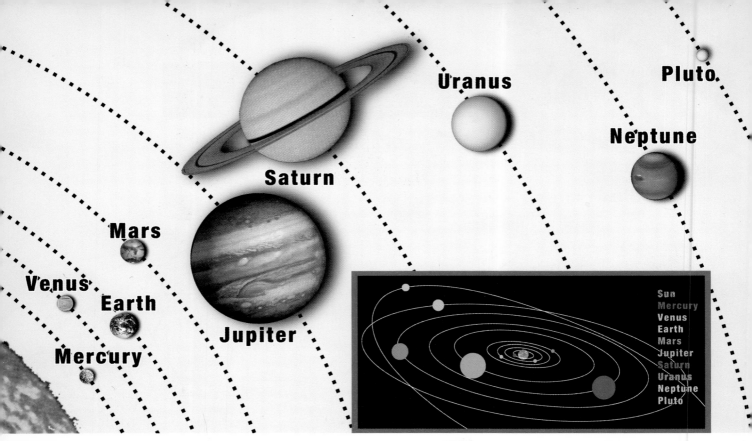

Saturn

Uranus

Pluto

Neptune

Mars

Venus

Earth

Jupiter

Mercury

Sun
Mercury
Venus
Earth
Mars
Jupiter
Saturn
Uranus
Neptune
Pluto

Earth Facts

Diameter–7,927 miles (12,756 kilometers)
Distance from Sun–93 million miles (150 million kilometers)
Moons–One
Revolution Period–365 days
Rotation Period–23 hours and 56 minutes

Earth and the Solar System

Earth is part of the solar system. The solar system includes the Sun, planets, and objects traveling with them. The solar system is always moving.

The Sun is the center of the solar system. Everything in the solar system circles around the Sun. The Sun is a star. A star is a ball of very hot gases. Stars like the Sun give off heat and light.

There are nine known planets in our solar system. Planets are the nine heavenly bodies that circle the Sun. Earth is one of them.

Earth is the third planet from the Sun. It is about 93 million miles (150 million kilometers) away from the Sun. Earth travels at the speed of 67,000 miles (107,200 kilometers) per hour. Earth rotates as it travels, too. This means it spins.

People live on planet Earth. They do not notice that it speeds through outer space. This is space outside of Earth's atmosphere. Atmosphere is the mix of gases that surrounds some planets.

Planet Earth

Early scientists did not understand Earth. They thought it was the center of everything. Early scientists thought the Sun circled around Earth.

The Polish astronomer Nicholas Copernicus (1473-1543) learned the truth. An astronomer is a person who studies stars, planets, and space. Copernicus discovered that the solar system's center is the Sun. Earth circles around the Sun.

Scientists and people also thought Earth was flat. They believed people could fall off Earth's edge.

Galileo Galilei (1564-1642) was another famous astronomer. He looked at the sky through his telescope. A telescope makes faraway objects look larger and closer. He discovered that planets are not flat. They are round like balls.

Scientists have taken pictures of Earth. The pictures show that Earth looks like a beautiful blue and white ball.

Pictures of Earth show that it looks like a blue and white ball.

Blue Planet

Earth is sometimes called the blue planet. Most of Earth looks blue from outer space. The blue color is from Earth's oceans and atmosphere. Oceans cover 70 percent of Earth's surface. Land covers the other 30 percent of Earth.

Water makes Earth different from other planets. No other planet has large amounts of liquid water. Earth is the only planet known to have life. Living things need liquid water to stay alive.

Other planets are too hot or too cold for liquid water. Then water takes on other forms. Sometimes it turns into a gas called water vapor. Other times it turns into a solid called ice. Only Earth's temperature allows all forms of water to exist. Temperature is the amount of hot or cold in something.

Most of Earth looks blue from outer space.

Atmosphere

Earth's atmosphere lies above its surface. The atmosphere acts like a blanket. Life on Earth would die without the atmosphere. The atmosphere protects Earth. It gives living things air to breathe.

Earth's atmosphere is made of gases and dust. It also has some water. Nitrogen makes up 78 percent of the atmosphere. Nitrogen is a gas that plants must have to survive.

Oxygen makes up 21 percent of the atmosphere. Humans and animals breathe oxygen. The atmosphere also has a special kind of oxygen called ozone.

Ozone forms a layer in Earth's atmosphere. It is called the ozone layer. It blocks out most of the Sun's harmful rays. Pollution hurts the ozone layer. People are trying to protect the ozone. New laws keep people from creating pollution.

Earth's atmosphere gives plants the nitrogen they need to live.

Parts of Earth

Earth is made up of three layers. The first layer is the core. The core is Earth's center. It is 2,200 miles (3,520 kilometers) thick. It is made of very hot rock and metal. Some of the rock has melted because of the heat. Liquid rock is called lava.

The second layer is the mantle. It surrounds the core. The mantle is about 1,800 miles (2,880 kilometers) thick. Heavy rocks make up the mantle. Some rocks are partly melted by the heat.

The third layer is the crust. This outer part of Earth is made of lighter rock. The crust is five to 25 miles (eight to 400 kilometers) thick. People live on top of Earth's crust.

The crust is broken into pieces called plates. Plates slide on melted mantle rock. Plates move very slowly. Continents are on plates. Continents are the seven large land groups of Earth. The plates are moving today, too. This movement is called continental drift.

Heat melts some mantle rocks into lava.

Continental Drift

Once, all of the continents fit together. But continental drift moved them apart. Continental drift is an important force. Moving plates change Earth's crust.

Sometimes two plates crash into one another. The plates push up rock. This forms mountains. Today, two crashing plates are making the Himalayan mountains. Every year the mountains become a little larger.

The movement of the plates also cracks Earth's crust. Volcanoes form along these cracks. Sometimes volcanoes erupt. This happens when lava comes through the hole in the crust. The lava is from the mantle layer.

Sometimes two plates have problems moving past one another. Their edges bump and rub against one another. This causes earthquakes. Two rubbing plates cause the earthquakes in southern California.

Two crashing plates are forming the Himalayan mountains.

Magnetic Field

Earth is like a large magnet. A magnet is a piece of metal. It pulls other metal toward it.

Melted metal makes up part of Earth's core. Earth's spinning makes the liquid metal flow. This moving metal creates a magnetic field. A field is a force that surrounds magnets. The field affects other metals. It attracts some metals. It pushes other metals away.

Earth's magnetic field helps keep the planet safe. It serves as a shield. It works with the ozone layer to block the Sun's harmful rays.

Sometimes the magnetic field traps tiny bits of the Sun. These bits can make Earth's gases glow. The glowing gases can make a colorful band of light called an aurora.

People can see auroras if they are close to Earth's poles. Poles are the most northern part and southern part of Earth. Sometimes auroras are also called northern and southern lights.

People can see auroras at the Earth's poles.

Rotation and Orbit

Earth is always spinning. It spins like a large top. It takes 24 hours for it to spin around once. One complete rotation equals one 24-hour day.

Earth's spin makes day and night, too. During daytime, Earth's spin turns half of the planet toward the Sun. Then that half of Earth faces the Sun.

While it is daytime on half of Earth, it is night on the other half. Earth's spin has turned the other half of the planet away from the Sun. During nighttime, half the planet faces the darkness of outer space.

Earth is always traveling around the Sun, too. It travels in a path called an orbit. Earth travels around the Sun once every 365 days. This is why each year on Earth is 365 days long.

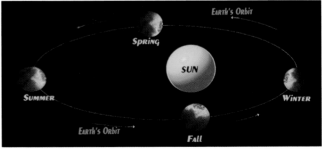

It is daytime when Earth spins toward the Sun.

The Moon

The Moon is Earth's only natural satellite. A satellite is an object that circles another object.

The Moon is one-fourth the size of Earth. Both Earth and the Moon have gravity. Gravity is a force that pulls things down. It keeps things from floating off into space.

Sometimes the Moon seems to change shape. But the Moon does not really change shape. The Sun lights different parts of it. From Earth, people can see only the parts of the Moon that the Sun lights. The part that people can see is called a phase.

People have wanted to search outer space for many years. Today, scientists are doing this. People have landed on the Moon. Robot-operated spaceships visit other planets in the solar system, too. This helps people learn about both Earth and the solar system.

The Moon circles Earth.

Hands On: Make a Rocket

Scientists build spaceships to search space. They move spaceships into space with rockets. You can make a simple rocket, too.

What You Need

A long (not round) balloon
Tape
A straw
A long piece of string

What You Do

1. Blow up the balloon. Tie the balloon's end so air cannot escape. The balloon is like a rocket.
2. Cut two pieces of tape. Use the pieces to tape the straw to the balloon. The straw is like a spaceship.
3. Tie one end of the string to something high.
4. Thread the other end of the string through the straw.
5. Tie the loose end of the string to something low.
6. Cut the the knot you used to tie the end of the balloon. Let the air escape.

Rockets work the same way as the balloon works. The balloon carries the straw along the string. Rockets carry spaceships into outer space.

Words to Know

astronomer (uh-STRON-uh-mur)—a person who studies stars, planets, and space

atmosphere (AT-muhss-fihr)—the mix of gases that surrounds some planets

core (KOR)—the inner part of Earth that is made of metal, rocks, and melted rock

crust (KRUHST)—the outer part of Earth that is made of lighter rocks

mantle (MAN-tuhl)—the layer of rock that surrounds the core

ozone (OH-zone)—a special kind of oxygen that blocks out some of the Sun's harmful rays

plate (PLAYT)—piece of the crust and mantle

Read More

Lauber, Patricia. *Seeing Earth from Space*. New York: Orchard Books, 1990.

Ride, Sally and Tam O'Shaughnessy. *The Third Planet: Exploring the Earth from Space*. New York: Crown Publishers, 1994.

Simon, Seymour. *Earth: Our Planet in Space*. New York: Four Winds Publishing Company, 1984.

Useful Addresses

NASA Headquarters
300 E Street SW
Washington, DC 20546

National Air & Space Museum
Smithsonian Institution
Washington, DC 20560

Internet Sites

Kids Web-Astronomy and Space
http://www.npac.syr.edu/textbook/kidsweb/astronomy.html

NASA Homepage
http://www.nasa.gov/NASA_homepage.html

StarChild: A Learning Center for Young Astronomers
http://heasarc.gsfc.nasa.gov/docs/StarChild/StarChild.html

Index